Automotive Industry Transformation

Virtual Reality in Vehicle Design

Table of Contents

Chapter 1. Introduction

Special Report: Automotive Industry Transformation - Embracing Virtual Reality in Vehicle Design

Navigating the automotive industry's transformation can feel like a thrilling, high-octane ride, particularly as cutting-edge technologies such as virtual reality (VR) revolutionize vehicle design. This Special Report aims to serve as your seatbelt on this wild ride, maintaining a grounded, simple explanation of the technical details, while illuminating the exciting implications VR holds for the industry and consumers alike. As you read through, expect to gain a robust grasp of how VR integrates into the lengthy and complex process of vehicle design—insights to propel you ahead in this growing field. Buckle up for a comprehensive view of the future of vehicle design that hinges on the digital realm, a potential game-changer for car manufacturers, designers, and ultimately, the drivers. Excited? Let's jump in!

Chapter 2. Introduction to Virtual Reality in Automotive Design

The automotive sector is standing on the brink of a design revolution, and the harbinger of this change is Virtual Reality (VR). This immersive technology is positioned to reinvent the traditional processes involved in automotive design, taking a deep dive towards a digital transformation that promises to reshape vehicle manufacturing processes, improve efficiencies, and redefine the end-user experience.

2.1. Evolution of Virtual Reality

Before we delve into the application of VR in the automotive industry, let's take a step back to lend some context to its evolution. The term 'virtual reality' was coined in the 1980s. However, the concept can be traced back to the 1950s and Morton Heilig's Sensorama, which was one of the first instances of multi-sensory, immersive experiences. Transitioning from these antecedents, VR today has morphed dramatically from simple projects simulating 3D worlds into advanced simulations driven by movement and gesture controls.

In-step with this evolution of VR technology, many industries, from entertainment to healthcare, began to adopt and refine the application of this technology. The automotive industry, in particular, has found compelling uses for VR through iterations and enhancements of the tech specifications, strengthening the industry's increasing reliance on this advanced tool.

2.2. Virtual Reality in Automotive Design

The application of VR technology across the automotive design process is indeed a game-changer. It offers immense potential for cost savings, efficiency optimization, and a significant reduction in the turnaround time in the design process while enhancing consumer experience.

Traditionally, the process of vehicle design had a long gestation period, involving multiple iterations, physical prototyping, tests, rigorous homologation processes, and eventual production. However, the adoption of VR technology offers an opportunity to bypass many of these laborious and expensive steps. For instance, VR allows for intricate simulations where designers can test different aspects of the vehicle design within a virtual environment before it hits the production floor. This process radically cuts the need for multiple physical prototypes and complex test setups, encouraging higher efficiency and delivering quicker results.

2.3. Immersion and Interaction

Immersion and interaction are the two fundamental principles that define virtual reality, and both lend themselves remarkably to automotive design.

- *Immersion*: VR technology completely immerses its users in a fully interactive and believable simulation of a 3D environment. This immersion becomes particularly relevant in automotive design, where designers can interact with a full-scale 3D model of a vehicle design, walking around and inside it to understand the design better.

- *Interaction*: VR's interaction capability allows designers to make changes to the model in real time, providing an unparalleled ease of tweaking and tuning aspects that were otherwise causing design limitations.

2.4. Flipping the Design Process

Using VR, designers can create vehicle models from scratch within the virtual environment. Changes such as modifications in the body shape, interiors, or even colour options can be executed and reviewed instantaneously. Vehicle dynamics, down to specific factors like steering feedback, suspension response, and ride quality, can be felt and tuned in real-time within this virtual world.

Simply put, VR significantly flips the traditional vehicle design process from one that was largely 'reactive' to a more 'proactive' process. The cost and time implications of every design iteration or modification made to the vehicle can be assessed instantly.

2.5. End-user Experience

End-users, a crucial link in the automotive value chain, also stand to gain massively from VR technology. It can be used to create virtual showrooms, allowing potential buyers to 'experience' the vehicle in a simulated real-world environment. Not just limited to exploring the car looks, the consumers can also experience various features and capabilities of the car in a way no brochure or presentation can capture.

In conclusion, the versatility of VR technology in the automotive design process comes with a promise to unravel innovative possibilities, disrupt traditional methodologies, and offer efficiencies. As we move forward, it is bound to bring an exciting change in the dynamics of the automotive design industry, providing a route to a future that was hitherto imagined only in the realm of science fiction.

Chapter 3. The Mechanics of VR and its Application in the Automotive Industry

The journey of integrating VR in automobile design commences with an understanding of the mechanics involved. The core of Virtual Reality (VR) involves three dimensions: Simulation, Interaction, and Communication - each interconnected to provide immersive, realistic experiences. Let's delve into each of these aspects comprehensively.

3.1. Simulation in VR

Simulation is the crux of VR, providing a virtual environment that can mimic the real world or manifest imaginary scenarios. VR employs exponential technologies like computer graphics, multimedia computing, and artificial intelligence to synthesize complex, lifelike scenarios.

Specifically, within the auto industry, simulations enable designers to model cars digitally. By incorporating advanced physics engines, these models accurately mimic real-world vehicle behavior. Designers leverage these simulations to test a myriad of elements like fuel efficiency, aerodynamics, safety features, and overall performance. This dramatically cuts down physical prototyping costs, paving the way for expedited, cost-efficient design processes.

3.2. Interaction with VR Environments

With VR, interaction is not limited to clicking buttons on a 2D screen. It incorporates 3D space, offering a completely immersive experience. VR environments are manipulated via various techniques such as gaze tracking, hand gestures, voice commands, and haptic feedback. This leap in interactivity echoes the design processes used in the automotive industry.

Designers and engineers can manipulate car models with intuitive hand gestures, tweaking designs real-time. Combined with the power of haptic feedback, they can 'feel' the impact of their iterations. This robust sensory feedback in VR environments, coupled with integration of accessories like VR gloves and suits, intensifies the illusion of interacting with a physical car model.

3.3. Communication within the VR Environment

In its essence, VR is an immersive communication medium. The automotive industry uses this potential to enhance collaborations during the design process. Experts located globally can now participate real-time in vehicle design, contributing insights and alterations as if they were physically present in a single design studio.

Internally, the synchronization of data ensures all team members review identical versions of design models. Externally, VR aids demonstrations to clients or stakeholders, providing a firsthand experience even before the car hits the manufacturing line.

3.4. Applications of VR in the Automotive Industry

The automotive industry has been astute in recognizing the potential of VR and adopting it in various domains.

3.5. VR in Vehicle Design

Vehicle designing is a meticulously iterative process, and VR brings enormous improvements to this procedure. Designers can create, modify, and finalize models within the VR environment, experimenting with different parts, colors, and finishes. This digital method eliminates the requirement of physical scale models, pushing towards a greener environment and substantial savings in resources.

3.6. VR in Prototyping and Testing

Through VR, full-scale, drivable models can be prototyped to test functionality and safety measures under various conditions. Crash tests, aerodynamics, fuel efficiency are evaluated in the VR space, reducing the risks and costs associated with real-world testing. There are additional considerable environmental benefits: reduced waste and carbon footprint.

3.7. VR in Manufacturing

Major automakers use VR foolproofing methods to reduce potential errors during production. These companies use VR to observe manufacturing processes, verify assembly sequences, and rectify the potential bottlenecks before the expensive physical assembly begins.

3.8. VR in Marketing and Sales

Finally, VR is rapidly gaining traction in automotive marketing. Dealerships now offer virtual showrooms where customers can inspect vehicles in a 360-degree view, customize features and take virtual test drives - creating an engaging, user-centric approach to car sales.

In summary, VR is an exceedingly promising technological marvel that holds the potential to transform the automotive industry's landscape. The leap from a 2D screen to a 3D virtual space has begun, and the ride is nothing short of thrilling. With this comprehensive understanding of the mechanics of VR, its application spaces in the car industry, and ascension towards a digitally-driven future, it's safe to say that the revolution is on the highway and gaining speed. For the automotive sector, integrating and evolving with VR isn't just a choice; it is a route towards unfathomable possibilities and a greener, more efficient future.

Chapter 4. The Evolution of Vehicle Design: From Physical to Digital

In the annals of human history, the invention and evolution of vehicles remains one of the most prominent features. Wheeling through the ages, vehicle design has gone from rudimentary, hands-on sketches and models to sophisticated, intricate computer-assisted design techniques. Today, we stand on the precipice of another monumental shift—from the physical domain of conventional design practices to the boundless digital world.

4.1. Beginnings of Vehicle Design

The dawn of vehicle design began in the confines of manual labor and rudimentary tools. In the early 20th century, draftsman artistry and craftsmanship ruled supreme, with every car design emerging from pencils, chalks, and paper. Each prototype was essentially hand-crafted—a labor-intensive, time-consuming process that required great skill and even more patience.

As automobile manufacturers grew and competition intensified, the need to innovate and produce more sophisticated designs in a shorter time span became apparent. The answer came in the form of clay modeling—still an analog approach but offering more flexibility and capacity for iterative design and refinement than earlier methods. By mid-century, clay modeling became a standard in automobile design studios.

4.2. Transition to Computer-Aided Design

The 1970s marked a turning point in the industry as computer-aided design (CAD) began to emerge. The advent of CAD software revolutionized the design field, offering a level of precision and ease of manipulation previously unimaginable.

With the ability to create complex, three-dimensional designs, test them for structural and aerodynamic efficiency, and alter them quickly and seamlessly, CAD became indispensable. Programs like AutoCAD became the go-to tools for vehicle designers, accelerating the process and eliminating numerous physical modeling steps.

However, CAD also came with its challenges. While it empowered highly accurate 3D design generation, the technology didn't initially lend itself to user-friendly interaction or the ability to immerse a designer within the design space. It still necessitated training and practice to master the software, and the rendering was limited and somewhat disjointed from the human understanding of spatial relationships.

4.3. The Digital Transformation: Introduction of Virtual Reality

With the dawn of the 21st century, the industry saw another technological shift—Virtual Reality (VR) was introduced into the design process. VR, with its interactive and immersive environment, perfectly melded the precision and flexibility of digital design with the spatial understanding inherent in physical design.

Vehicle design was no longer limited to the flat screen. Instead, designers could now literally walk around, interact with, and view their creations from every angle. Changes could be visualized instantly, facilitating immediate feedback and reducing the need for multiple physical prototypes.

One key advantage VR provided was the ability to see the design in a way that replicated real-world perspectives. Designers can virtually sit inside their designs, gauging ergonomics and visibility. They can "feel" the car, manipulating its components and dimensions, evaluating whether everything fits and works together in harmony.

Although the adoption of VR was initially slow, due to cost and availability of technology, it has rapidly proliferated in the last decade. Today, every major automobile manufacturer employs VR technology extensively in the vehicle design process.

4.4. The Digital Future: AI and VR in Vehicle Design

While VR has undeniably revolutionized today's vehicle design, its future potential in combination with other emerging technologies is even more exciting. Artificial Intelligence (AI) is rapidly converging with VR. When applied to vehicle design, AI can learn from millions of data points provided by users and vehicles, then provide designers with optimized suggestions for modifications and improvements. Combining AI with immersive VR will allow designers to interact with these AI-generated design iterations in a hands-on, fluid manner. Design rounds will be more about selecting the best AI-proposed design rather than creating every detail from scratch.

Moreover, coupling VR with haptic feedback is another exciting development. This technology allows designers to 'feel' the texture and tactile experience of different materials, directly influencing the choice of materials used in the vehicle, thus enhancing the design further.

In conclusion, the application of VR in vehicle design is not just transformative—it's revolutionary. It has propelled vehicle design from the physical realm to the digital, expanding the boundaries of innovation drastically. As advancements in technology continue to arise, VR will further blur the line between the static screen and the interactive, all-around experience. In doing so, it transcends the traditional limitations of both the physical design board and the limitations of early digital designs, pioneering a new era in vehicle design, replete with possibilities as infinite as the digital realm itself.

Chapter 5. Exploring Commercial VR Platforms for Vehicle Design

Tech giants, start-ups, and independent developers alike are making giant strides to leverage virtual reality (VR) for commercial use, particularly in the automotive industry. This drive is underpinned by an array of commercial VR platforms, each boasting unique features that are reinventing the age-old process of vehicle design.

5.1. Understanding Key Players in Commercial VR

Tech juggernauts like Google, Microsoft, and Facebook have been making significant advances in VR technology. Google, for instance, has Google Cardboard and Daydream, two VR platforms that leverage the use of smartphones for a more accessible VR experience. These platforms provide a basic entry point to VR and have found utility in simple applications of vehicle design, such as the preliminary stages of concept development.

Microsoft's HoloLens, on the other hand, is a more advanced mixed reality platform. Its ability to insert CGIs into the user's real-world environment has proven revolutionary. Automotive companies can now project full-size 3D models of vehicles, allowing designers and engineers to interact with the vehicle design in a dynamic way that wasn't previously possible.

Facebook's Oculus Rift and Oculus Quest 2, which are high fidelity VR solutions, offer a deeply immersive experience, allowing for detailed interior and exterior design evaluations. With Oculus platforms, designers can simulate the driving experience, enabling a detailed analysis of cabin ergonomics before crafting any physical prototypes.

Beyond these mainstream platforms, numerous specialized VR solutions are designed to meet specific enterprise demands. Companies like Varjo tout 'human-eye resolution' VR and XR (extended reality) headsets, adept at supporting tasks in automotive design that demand high visual fidelity.

5.2. Leveraging Platform Strengths for Vehicle Design

Each VR platform has different strengths and weaknesses, and understanding these is crucial when choosing the right tool for specific design tasks. Basic platforms like Google Cardboard and Daydream provide broad strokes. They allow auto designers to share and view vehicle concepts easily, fostering quick feedback in the early rudimentary stages of design.

More advanced platforms like HoloLens and Oculus lend themselves to the more intricate parts of design. They enable designers to visualize and interact with the vehicle at scale, allowing them to assess and enhance elements like the vehicle's usability and aesthetic appeal.

Moreover, a notable aspect of using these advanced tools is the ability to check for design deficiencies and make necessary alterations in real-time. For instance, with Oculus, you can simulate different driving conditions and scenarios, which allows designers to make crucial adjustments to ensure optimal user comfort and safety.

5.3. Industry-Specific Platform Developments

Increasingly, big players in the automotive industry are investing in their proprietary VR solutions for vehicle design. Ford, for instance, uses a VR tech dubbed Ford immersive Vehicle Environment (FiVE) to review vehicle designs. Audi, too, employs Audi Virtual Reality Experience for design and configuration processes.

However, building an in-house VR solution is a colossal undertaking requiring significant resources. Many companies, especially smaller ones, opt for commercial platforms but tailor these to fit their specific needs. This mixture of proprietary and adapted software is the current state of VR incorporation in the critical process of vehicle design.

5.4. Gap Analysis and Future Developments

Despite the impressive developments and applications of VR in vehicle design, gaps remain that prompt ongoing innovation. For instance, while Oculus and Varjo provide high-definition experiences, they still fall short of delivering a fully realistic, high-resolution environment. Additionally, physical feedback, an integral part of vehicle design and evaluation, is limited in current VR capabilities.

To tackle these deficiencies, we can expect advancement in VR technology, such as haptic feedback wearables and vastly improved visual capabilities. Investments in machine learning and artificial intelligence will likely drive further improvements in VR platforms, making it possible for computers to anticipate user inputs and learn from user interactions.

5.5. The Direction of Commercial VR in Vehicle Design

Given the impact VR has had in a short span, adoption of VR in the automotive industry is inevitable. The use of VR platforms to design, prototype, and experience vehicles before they've been physically manufactured reduces overall production costs significantly and accelerates the design-to-market process.

The real challenge, though, resides not in whether to use VR, but in choosing the right VR platform for the job at hand. As each player in the commercial VR industry continually refines and enhances their offerings, car manufacturers must stay abreast of these developments in order to harness the full potential of VR for vehicle design.

Equally compelling is the growing trend towards tailoring widely-available commercial VR platforms for unique enterprise use. The melding of established VR software with tailor-made modifications allows for a customized workflow that is optimally suited to the specific demands of vehicle design.

In all, the future of automotive design lies in the interplay of VR technology and immersive experiences. That fusion, powered by commercial offerings and bespoke adaptations, is set to redefine innovation in the automotive industry, enabling car manufacturers, designers, and drivers to take part in a thrilling journey towards the future of automobile development.

Chapter 6. Industry Case Studies: VR in Renowned Auto Makers

Modern auto manufacturing and design no longer look to be limited to 3D models and hands-on tinkering. A glimpse into their current operations reveals a surprising affinity for virtual reality (VR).

6.1. The Innovation Engine: Ford

Ford is a key player in the adoption of VR technology in the automobile industry. Their Immersion Lab in Dearborn, Michigan, is a testament to this shift. Inside this facility, teams of engineers and product designers review 3D car models using VR headsets. Ford uses this immersive technology to assess vehicle attributes that would typically require physical prototype reviews. Everything from visibility, reachability, spaciousness, and even the texture of materials is evaluated in this virtual environment.

These VR sessions also offer a global collaborative platform. Designers and engineers from Ford's global operations can interact in real time, making model adjustments while discussing potential improvements. This approach drastically reduces cycle times and travel expenses.

Thus, Ford's VR application not only streamlines the design process but also increases the efficiency in prototype development and promotes global collaboration, mitigating the barrier of geographical distance.

6.2. The Visionary: Volkswagen

Volkswagen, another exemplar in the automotive sector, has shown a similar enthusiasm for VR. Volkswagen's Virtual Engineering Lab in Wolfsburg has seen its team use VR to design vehicles in a 'gravity-free' virtual environment.

Teams of designers and engineers wear VR glasses and use voice commands to navigate the virtual environment, manipulating 3D models of vehicles. Notably, auto parts that typically weigh several kilograms in real-life can be easily manipulated with a simple hand gesture in the virtual world, accelerating the design process.

Furthermore, Volkswagen introduced the 'Digital Reality Hub' in 2017. This hub is a collaborative VR platform that enables cross-functional and cross-plant team collaboration. This allows experts from different disciplines and geographical locations to work together on a single platform, demonstrating how VR enables synchronous global collaboration in the design process.

6.3. The Game-Changer: Audi

Audi stands out with its unique VR applications in the design process. They gave a significant makeover to their traditional clay modeling design process by introducing pre-series prototypes into the virtual environment.

'Audi's Virtual Reality Software', AVRES for short, in the Concept Design Center in Beijing uses VR to create full-scale models of upcoming vehicles. The relief of the car body, the proportions, the size effect, and the light and shade - every aspect of the vehicle can be experienced realistically in this virtual world. Furthermore, AVRES enables multi-user sessions, allowing several employees to inspect the model simultaneously and propose changes in real-time.

Audi cultivates a continuous refinement process, where the virtual car models evolve gradually in the VR environment until they meet the desired design standards. This echoes the traditional iterative clay modeling process, but with the added benefits of increased flexibility, time efficiency and cost savings.

6.4. Looking Ahead

Virtual Reality is expanding its sphere of influence, and the automotive industry is undergoing a transformation in design norms. It is harnessing the benefits of VR for better collaboration, efficient design processes, and tangible cost savings. Seeing the success of these industry giants, it's perhaps just a matter of time before VR becomes the norm in vehicle design, bringing an exciting future nearer to reality than ever imagined.

Chapter 7. The Impact of VR on Speed and Efficiency in Design

The advent of Virtual Reality (VR) technology marks a profound shift in the way vehicles are designed, elevating the process from the tangible, physical domain to an immersive 3D virtual environment. Speed and efficiency in design, two factors paramount for manufacturers, witness significant gains as VR reduces the lead time for designs, streamlines the process, and minimizes errors.

7.1. Integrating VR into the Design Process

Traditional vehicle design process involves preliminary sketches and 2D cad modelling, undergoing hydraulic test analysis, and then producing a 3D physical prototype. This chronological path is time-consuming and requires considerable physical resources.

Integrating VR into the vehicle design process enables designers to bypass several steps. An initial 2D sketch or CAD model can directly be turned into a 3D VR model and inspected thoroughly before undertaking any physical prototyping. From initial concept to final design, working within a virtual environment enables designers to visualize the final product, adjust or refine designs in real time, and isolate potential disadvantages at an early stage.

Designers can also easily switch between the minutiae of parts at the component level to the holistic view of a complete car via the capabilities of VR. From the contours of the exterior body to the wiring inside the dashboard, VR allows detailed inspection of every component individually as well as in the context of the entire vehicle.

7.2. Impact on Design Speed

When it comes to the automotive design cycle, time is expensive. A traditional approach can take several weeks, even months to produce a physical prototype. VR can trim those weeks down to days. The lead time in manufacturing prototypes gets substantially reduced with VR because you can attain design approvals virtually.

It also puts an end to geographical constraints in design collaboration. Designers from disparate locations can simultaneously work on the same project. With the advent of cloud-based VR, global design teams can not only share, modify, and approve designs in real time—but can also do so from the comfort of their offices.

7.3. Effect on Design Efficiency

Equally significant as the speed of the design process is its efficiency. The automotive industry spends billions annually on prototypes—only a small percentage of which end up going to manufacturing.

Through VR, designers can now step inside the car design, review it, and make changes in real time, without having to produce an actual prototype. This process drastically reduces the material and resources involved in physical prototyping—making design iterations more cost and time efficient. Moreover, stimulating user interaction with VR fosters a more intuitive feeling of the design, which helps to uncover issues that might not be apparent in a traditional 2D representation.

7.4. Enhancing Error Detection and Mitigation

The primary advantage of VR lies in its ability to simulate real-world driving conditions and experiences. This futuristic model facilitates conducting virtual tests such as aerodynamics and crash tests, generating invaluable data regarding the performance of the design under various conditions.

Identifying design issues early can save significant time and money that would otherwise be invested in manufacturing, testing, and subsequent iterations of physical prototypes. VR helps catch spatial problems, component clashes, and assembly issues that might be overlooked in less detailed design reviews. Isolating these problems before the actual manufacturing phase prevents costly redesigns further along in the development process.

The advent of VR also provides designers with the opportunity to perform ergonomic evaluations. Ergonomics plays a vital role in user comfort, safety, and overall driving experience. With VR, designers can analyze and predict user interaction and comfort based on the interior design of the vehicle. This ability to conduct ergonomic analysis in the virtual world ensures the vehicle offers optimal operational comfort—without the production of a single physical prototype.

7.5. Bridging the Gap between Designers and Clients

One of the main challenges of classic design methods is conveying the final design to clients. VR simplifies this process by providing an interactive, 3D visualization of the design. Clients can virtually experience the vehicle, interacting with it as they would in real life. This approach eliminates ambiguity and ensures that the client has a clear and comprehensive understanding of the end product.

7.6. Conclusion

The arrival of VR has set the automotive industry on a new trajectory of growth and innovation. While traditional design processes are laborious, time-consuming, and result in extensive resource expenditure, VR offers a highly efficient, cost-effective alternative.

With an immersive 3D design platform, accelerated design phases, minimized error frequencies, and enhanced client-designer interactions, VR significantly elevates the speed and efficiency of vehicle design.

Virtual Reality has the potential to streamline the current design process within the automotive industry, creating an agile, efficient, and future-ready structure. Its influence extends from the drawing board to the manufacturing floor, pointing towards an exciting direction for the industry as it accelerates towards a digital future.

Chapter 8. Confronting Challenges: VR Integration and Skill Development

Despite the shining promises VR technology holds, its integration into the automotive industry, especially for vehicle design, is not clear-cut. Several hurdles need to be recognized, understood, and overcome for it to unlock its full potential. This section delves deep into these challenges: from technical aspects to skill-set development obstacles.

8.1. Technical Challenges of VR Integration

The technical roadblocks to VR integration in the vehicle design process are multifaceted. VR's infrastructural demands, such as high-performance hardware and advanced software, post considerable hurdles. Only the best graphics cards can run VR applications smoothly, and affordable solutions in this regard are few and far between.

Moreover, automakers are required to invest significantly in up-to-date VR systems and subsequent upgrades, putting immense pressure on budgetary allocations. Add to this the issues of latency and motion sickness that some users experience with VR, and the picture only gets murkier. Lagging visuals can greatly impact the design process and turn an otherwise interactive tool into a disjointed, inconvenient liability.

Despite these challenges, it is undeniable that technological advancements are exponentially erasing some of these hurdles. Nevertheless, the speed of such progress is a critical factor to consider, keeping in mind that adaptability within the industry is not always prompt.

8.2. Skill Development in a VR Environment

Incorporating VR into the automotive workflow is not just a matter of providing the necessary technical infrastructure; it also needs a workforce adept at using these advanced tools. As a relatively new technology in the sector, VR lacks a pool of skilled users and thus demands significant investment in training and skill development.

Not all design engineers have a grasp on the VR technology, which has a steep learning curve. Mastering the skills to optimize the use of VR tools takes considerable time and can slow the pace of design and production initially. This calls for a comprehensive skill development framework that allows for the seamless transition of employees into the digital realm of vehicle design.

8.3. Bridging the Skill Gap

Crafting a tailor-made strategy for skill development begins with understanding the breadth and depth of the required skill set. VR in automotive design is not just about grasping the operation of hardware and software. It also includes building a strong foundation in 3D design principles, understanding the specifications of VR modeling, and honing skills in rapid virtual prototyping.

Ongoing training programs and workshops can help employees navigate these new territories. In-house training is essential and can leverage experienced staff to mentor those still learning. Additionally, collaborations with leading VR firms can provide unique insight into best practices and emerging trends, keeping the design teams on the cusp of innovation.

8.4. Overcoming Resistance to Change

Adoption of novel technology often meets with resistance, largely due to apprehension about change. The radical shift from traditional design processes to a VR-based workflow may unsettle many designers. This calls for a proactive change management strategy that fosters a culture embracing transformation.

User-friendly incorporation of VR systems, paired with comprehensive training, is essential to tackle the technical aspects of this resistance. However, overcoming psychological barriers might be a sturdy challenge. Convincing people of the value of change, demonstrating the benefits to the individual and the company, and ultimately working towards a VR-acclimatized workforce will require persistent efforts. Automakers must take care to adopt a gradual transition process rather than making sudden, workplace-shifting changes.

8.5. Future-proofing the Automotive Industry

To make the integration of VR into vehicle design a lasting success, automakers must future-proof their strategies to accommodate the evolution of technology. This involves adaptable institutional policies, upskilling agendas, and a responsive infrastructure that evolves with the technology.

Staying up-to-date with VR advancements and ongoing research could provide the necessary impetus for making informed decisions. Companies should foster a spirit of innovation, encouraging designers to experiment with new VR features and techniques. This way, each stride in VR technology could be transformed into a potential leap forward in vehicle design.

Undeniably, integrating VR into vehicle design comes with its share of challenges. However, with the right strategies and a well-crafted change management plan, these hurdles can be tackled effectively. Making VR a staple in vehicle design might take time and patience, but the benefits it promises are unparalleled. This makes confronting and surmounting these challenges an exciting prospect for the automotive industry.

Indeed, riding the wave of VR technology is no easy task. Still, with an able crew trained to navigate through the storm, the automotive industry can emerge more robust, innovative, and ready to deliver cutting-edge vehicles that push the boundaries of design and deliver a superior experience to the end consumer.

Chapter 9. Implications of VR for Consumer Car Customization

The future of the automotive industry stretches far beyond mere aesthetics and performance. Technologies such as artificial intelligence, the internet of things, and importantly for this discussion, virtual reality (VR), are reshaping every aspect of vehicle production, distribution, and use. Among the many potential applications of VR technology, one of the most promising lies in consumer car customization.

The concept of customization has always intrigued car enthusiasts, allowing them to express their personality and preferences through their vehicles. Traditional customization used to be a physical, manual process—changing car paints, swapping rims, adding spoilers and more. Today, digitalization and the use of VR, in particular, offers an opportunity to accelerate this process, adding a level of personalization that simply wasn't possible before.

9.1. Transformation of the customization process

Thanks to VR, the customization process can occur much before a physical copy of the vehicle even exists. Using a VR headset, potential car owners can experiment with various customization options, visualizing their modifications in real-time in a three-dimensional space. This adds another dimension to the customization process, providing an experience more immersive than a flat computer screen or a sculpted clay model.

Three-dimensional rendering in a virtual environment allows users to view their chosen vehicle from all angles. They can zoom in for a close-up regard of specific parts or get a bird's-eye view of the vehicle. They can tweak colors, shapes, and materials of virtually every single part, whether it's the exteriors like the body paint, rims, or spoilers, or the car interiors including the upholstery, dashboard, and infotainment system. This level of detail ensures that the car owners get precisely what they envisage.

9.2. Wider reach, inclusivity, and massive savings

Along with transforming the process itself, VR enhances the outreach of customization. Individuals can customize their cars from the comfort of the home without needing to visit a dealership. This convenience also comes with significant cost savings, reducing the manual labor required and consequently the expenses associated with it.

Additionally, potential car owners can simulate the performance of the new vehicle using VR. As they make modifications, the system recalculates and presents the new performance parameters like speed, fuel efficiency, or aerodynamics, allowing them to make an informed decision.

9.3. Seamless integration with other technologies

Another exciting potential of VR lies in its seamless integration with other emerging technologies. For instance, integration with artificial intelligence can create algorithms that learn from the customer's previous choices to suggest new customizations. Similarly, the amalgamation of VR with the Internet of Things (IoT) allows consumers to build a "digital twin" of their car and experiment with it in various settings and under varied conditions. This kind of multi-platform, interconnected technology is changing the way the world thinks about vehicle customization and ownership.

9.4. Challenges and the Path Forward

While VR offers an array of exciting opportunities to revolutionize car customization, there are certain challenges. The initial investment in these technologies isn't cheap. Creating VR models of vehicles with customizable features requires specialist knowledge and time. Encrypted data transfer mechanisms have to be in place to uphold security and to ensure that the customer's VR designs are safe, private, and tamper-proof.

Considering the tech-savviness required for VR usage, there might be a learning curve for some consumers. This necessitates strategizing user-friendly interfaces, tutorials, and customer assistance to ensure easy accessibility, even for those not familiar with VR.

Though it might take some time before we see full-fledged VR car customization becoming commonplace, the signs are quite promising. VR, in combination with other emerging technologies, holds the potential to transform the very essence of car ownership and experience.

In essence, VR technology is paving the way for an automotive future where consumers no longer have to be content with a handful of customization options selected by manufacturers. They can have an unprecedented level of involvement in shaping their vehicle, tailoring it to their needs and tastes. Let's watch the wheels spin as we steadily cruise towards this exciting future.

Chapter 10. Future Perspectives: How VR could Shape the Cars of Tomorrow

As we glimpse into the future with the lens of virtual reality, several striking possibilities come to light, radically reimagining the process of vehicle design and manufacturing. The immersive nature of VR paired with the breadth of its applications places it at the heart of the automotive industry's digitization crusade. From conceptualization to troubleshooting, virtual test drives to customer personalization, VR promises to reshape and streamline the way we craft, and consecutively, interact with vehicles.

10.1. Potential of VR in Conceptualization and Prototyping

Envision drafting a new car model, not with pencil and paper, but with a VR headset and handheld controls. The ability to sketch in three dimensions vastly expands designers' creative realm, allowing them to visualize and manipulate designs intuitively in actual space and scale. Designers can simulate various materials, tweak shapes and proportions, and view the vehicle from any perspective — a substantial upgrade from traditional 2D sketches and even sophisticated computer-aided design (CAD) software.

Beyond sketches, high-fidelity VR models eliminate the need for time-consuming and costly physical prototypes. Designers thoroughly inspect every element of a vehicle in interactive detail, many layers deep into the digital prototype, enabling early identification and rectification of design issues. Furthermore, VR incorporates haptic feedback technology, bridging the gap between the digital and tactile world, granting designers the ability to 'touch' and feel textures and surfaces while making virtual alterations. Naturally, this 3D prototyping process could result in time and cost savings, as well as heightened design precision, which the traditional cycle of sketching, prototyping and iterating often fails to deliver.

10.2. Virtual Test Drives - A New Reality

VR test drives stand as an illustrious possibility. While no VR simulation can fully replicate the physicality of driving, they offer an exceptional tool for engineering evaluation and a method of transporting potential buyers into a car's cockpit even before it hits the production line. Engineers could examine vehicle performance under virtual, yet highly lifelike, driving conditions, measuring acceleration, handling, and braking response, among other factors.

Customer test drives also undergo a remake. With VR, individuals touch, feel, and 'drive' prospective vehicles in varied and extreme settings without leaving the showroom. Such immersive test drives could unveil insights about the user's emotional responses to the automobile, a valuable addition to the typically analytical data sets gleaned from traditional test drives.

10.3. Realizing Customer-driven Personalization with VR

VR enables unprecedented customer personalization. Similar to configuring a virtual character in a video game, consumers could employ VR to customize their vehicle. Everything from color schemes, upholstery materials, accessories, and more, can be sampled and adjusted to personal preferences in the virtual world, before securing a locked-in design ahead of actual production. This poised leap from pre-set options to a full-fledged customization experience revitalizes the customer journey, potentially contributing to customer engagement and satisfaction.

10.4. Troubleshooting and Training

VR transcends current practices of troubleshooting, enabling technicians to simulate and study malfunctions in a virtual environment. They can peel away layers, replace parts, and investigate the inner mechanics without managing a physical car. This experience could both reduce the time it takes to identify the problem and minimize the potential for error.

Beyond diagnostics and repair, VR is set to be an influential tool for training future automotive engineers and technicians. Novices can learn and practice in a safe, controlled, virtual environment replete with real-time feedback, intricate simulations, and performance tracking, all while sparing the equipment wear and tear typical of traditional hands-on training.

10.5. Redefining Showrooms and Sales

The dawn of VR could disrupt traditional automotive sales models. Digital showrooms, accessible from the comfort of a customer's home, could showcase every model in a car company's lineup, a feat nearly impossible for most physical showrooms due to spatial limitations. Customers would be able to explore every inch of a vehicle, experience various color options, and enjoy immersive test drives in an assortment of environments. This could revolutionize the shopping experience, giving customers access to the complete fleet at their fingertips.

10.6. The Road Ahead

As thrilling as these possibilities sound, implementing VR in the automotive industry will not be a smooth drive. Challenges related to technology adoption, infrastructure upgrade, training needs, and even psychological aspects like motion sickness are substantial. But as the technology matures and adoption rates rise, the automotive industry could find itself poised on the brink of a new digital age, one that is more time and cost-effective, offers an enhanced user experience, and accelerates innovation.

Ultimately, the transformation towards VR in vehicle design and manufacturing requires a strong collaborative will among industry stakeholders. The journey will necessitate significant investments in technology, training, and change management. But for a future where design ideas come to life instantly, test drives occur without a physical car, and every customer gets to customize their vehicle to their heart's content, the return on investment is enormous.

As progress rockets ahead, it is clear that automotive professionals who adapt to the changing technological landscape stand to gain a competitive edge. For those willing to ride this daredevil wave of innovation, the future of car design powered by VR represents a road marked with promise and potential, leading us towards a future automotive industry that faces tomorrow's challenges head-on and with unwavering determination to revolutionize.

Chapter 11. Roundup: Key Takeaways and Industry Outlook

The potential influence, optimistically and disruptively, of virtual reality (VR) on the automotive industry is evident. Changes are already being felt, with several car manufacturers incorporating VR into their design process, with many more to join in the future. This shift towards integrating VR into vehicle design facilitates greater efficiency, creativity, and safety, all while potentially reducing costs and time-to-market for new vehicle models.

11.1. VR's Role in the Automotive Industry

Virtual reality is becoming an influential prerequisite tool for the automotive industry, primarily utilized in vehicle design and conceptualization. Car manufacturers are using VR to foster an environment that encourages creativity, reinforces safe testing measures, consolidates design iterations, and reduces prototyping expenses.

With VR, manufacturers can design cars and render them in a virtual space accessible from anywhere across the globe. As a result, different teams can contribute to the design process simultaneously without having to be physically present in the same location. This worldwide collaboration not only hastens the design process but also harbors diverse ideas from various design centers globally.

Furthermore, VR offers a fail-safe environment for testing vehicle designs without endangering safety protocols. Risky tests like crash simulations can be carried out exhaustively at minimal expense, mitigating the risks traditionally associated with physical testing procedures.

Assessing the impact of design tweaks and iterations in real-time is another key advantage of using VR in vehicle design. Transparency is achieved as all the alterations are instantaneously visible to team members, fostering a shared understanding among all parties involved and maintaining design alignment.

11.2. Implications for Car Manufacturers

For car manufacturers, this digital shift implies significant capital-intensive initial setup costs. But the long-term benefits like improved design efficiency, cost savings in prototyping, and reduced time-to-market overwhelmingly justify these costs.

VR facilitates a smoother, more efficient vehicle design process. 3D CAD models of vehicle parts can be imported into virtual reality simulations, enabling designers to refine their models with greater precision and ease than ever before.

By minimizing the need for physical prototypes, companies can save on material and fabrication costs, and reduce their environmental footprint. Additionally, the capacity to conduct crash tests in a virtual environment significantly reduces the associated physical and fiscal risks.

Furthermore, the global connectivity that VR brings allows companies to tap into talent from across the globe, therefore benefiting from diverse perspectives and expertise.

11.3. Outlook for Car Designers

Car designers are required to familiarize themselves with VR tools and software to keep pace with the transforming automotive industry. While it presents a learning curve for many, the adaptability to new design tools is essential for sustained success.

Moreover, the transition to VR offers designers the potential to explore bolder, more original designs. Unbound by the practical constraints of the physical world, they can manipulate scale, color, and material in ways that were previously unthinkable.

Fostering a seamless link between digital design and physical realization is another key challenge for designers. They must find the balance between using VR to push design boundaries and ensuring that these designs are still practical and manufacturable in the real world.

11.4. Prospects for End Consumers

This digital upheaval within the automotive industry has encouraging implications for end consumers. VR allows for tighter feedback loops between manufacturers and customers, leading to vehicles that better meet users' needs and expectations.

Car manufacturers are also using VR to engage customers in the purchasing process. They use VR headsets to provide customers with immersive test-drive experiences even before the car is physically produced.

Moreover, customers can customize their future cars in real-time – changing colors, upgrading features, and requesting modifications – all while observing the immediate impact of these changes on the car's design.

11.5. Industry Outlook

The rapid adaptation of VR in the automotive industry foreshadows more widespread use in the coming years. As the technology continues to improve and become more cost-effective, small and midsize manufacturers will also begin integrating VR into their design processes.

The benefits of VR's ability to streamline design and manufacturing processes, while facilitating global collaboration and reducing costs, will likely lead to its widespread adoption across the industry.

The next stage will be integrating VR with other emergent technologies such as artificial intelligence (AI) and the Internet of Things (IoT). AI could provide predictive modeling, and when combined with VR, could allow for even more robust design processes.

Ultimately, the main challenge will be to make this technology accessible to smaller manufacturers, who may lack the necessary resource pool, both in terms of capital and skill. Collaboration between tech providers, governments, and car manufacturers may be crucial to overcome this hurdle, with potential solutions including subsidies or shared virtual design centers.

To conclude, the adoption of VR in the automotive industry signifies the dawn of a new era in vehicle design. While the challenges are significant, the promise and potential held by this technology are immense. The industry outlook for VR integration in vehicle design and production is both highly promising and exciting, heralding a new age in the automotive industry.

www.ingramcontent.com/pod-product-compliance
Lightning Source LLC
Chambersburg PA
CBHW071042260726
48661CB00007B/3104